FRACKING C[APITALISM]
Action plans for the eco-so[cial]

A World to Win writing team:
Penny Cole
Matt Worsdale
Gerry Gold
Donald McQueen
Editor: Paul Feldman

© A World to Win

All rights reserved
The moral rights of the authors have been asserted

ISBN: 978-0-9523454-8-0

First published by A World to Win
November 2013
2nd Edition April 2014

Website: aworldtowin.net
Network forum: aworldtowin.ning.com
Email: info@aworldtowin.net
Telephone: 07871 745258
Facebook: A World to Win
Twitter: @aworldtowin

Contents

Introduction	5
1 The Great Shale Gas Deception	7
2 Planning system silences the voice of the community	16
3 Resistance movement threatens the frackers	22
4 Trapped by a corporate state web	26
5 The road to ecocidal suicide	33
6 It's a socio-eco crisis – time to act	39
7 Frack capitalism to build a sustainable future	43
An action plan for the UK – 2014	49
Notes	55

Introduction

The message has gone out to corporations everywhere: Britain is open for fracking. In response, campaign groups now exist the length and breadth of Britain in opposition to the plans to industrialise the countryside with tens of thousands of drilling sites. They are taking legal action, lobbying their representatives and protesting and occupying sites at considerable risk of police brutality.

But this grass roots movement is up against formidable adversaries. Corporations have the backing of the state and a public relations campaign led by the government is promoting the lie that fracking is safe and will lead to cheaper energy.

Yet public support for shale gas extraction continues to fall while backing for renewables grows. Government claims about jobs and lower gas prices are exposed for the grand deceptions they are. None of this will deter the Cameron government, however, which has thrown the weight of the state behind the frackers.

Substantial tax concessions for companies that take the plunge have been matched by a substantial revision of the planning system so that it is now heavily weighted in favour of the drilling firms. Early in 2014 local authorities were informed they could keep 100% of all business rates generated by fracking activity in their area, which is double the usual amount. Despite austerity and severe cuts in vital services, the government said it would make up the difference. When it comes to fracking, money is no object for a government whose energy policy is in total disarray. That was demonstrated when massive state subsidies were provided for Chinese and French corporations to entice them into building new nuclear power stations.

No surprise then when the France-based energy giant Total announced a major investment in UK fracking. Thwarted by a government relatively hostile to fracking at home, Total understands that the British government will not stand in its way. On the contrary, ministers spent the end of 2013 lobbying hard in Brussels as the European Union prepared to issue fairly tough regulations covering fracking. It paid off. In January 2014, the EU issued "recommendations" rather than regulations and glasses were raised in Whitehall.

The unholy alliance between the corporations and the state is openly acknowledged: As Dart Energy chief executive John McGoldrick admitted, "we have tremendous government support". Chapter 4 *Trapped by the corporate-state web*, reveals how the connections between lobbyists, former BP chief Lord Browne, government departments and ministers have produced results – for the industry. Browne, the chairman of fracking firm Cuadrilla, wants more. He believes extracting shale gas is a "national imperative" and bemoaned the fact that achieving planning permission still takes time.

He is kicking at an open door. Cameron told a news conference in late 2013: "On fracking, we do need to take action across the board to help enable this technology to go ahead. There is a worry people are going to have to go through so many different permits in order to start fracking that they simply won't bother, so we need a simplified system." What this implies is that the government could use its reserve powers to have planning applications for shale taken at ministerial level. This would cut out troublesome and slow-moving local authorities. Cameron is lobbying hard for fracking in the EU, and claims all it will take is to get a few wells up and running and all the opposition will disappear.

This unholy corporate-state alliance to impose fracking on communities has devastating consequences for the environment at local level and contributes to climate change, as we set out in Chapter 5 *The road to ecocidal suicide. Fracking Capitalism* shows how these considerations are secondary for an economic system driven primarily by profit and maximising shareholder value. The most recent (March, 2014) IPCC report on the impacts of climate change lists hunger, mass migration and extreme economic and social disruption as some of the outcomes of climate change that are imminent but governments ignore this evidence. That is why Chapter 7, *Frack capitalism to build a sustainable future*, argues: "Slowing, then halting, global warming requires an immediate reduction in the quantity of fossil fuels burned. The major question facing humanity is: can we rely on the existing governing structures to make this happen? The evidence against this is clear enough."

We set out a comprehensive plan of action to tackle what is in reality an eco-social crisis, where the political-state system is in cahoots with the corporations to frack us all. Going beyond building resistance to fracking to creating a movement that puts power firmly in the hands of presently powerless communities is what is proposed. A World to Win invites you to join with us in carrying this project through to success.

Paul Feldman
A World to Win Communications Editor
April 2014

1
The great shale gas deception

Widespread hydraulic fracturing – fracking – for shale gas threatens large areas of Britain with unimaginable industrialisation and pollution of air and water.

The dash for gas is sponsored and driven on by the state, which holds out the prospect of cheaper energy as well as many more local jobs. Both are deceptions on a grand scale by a system that has wilfully turned its back on carbon-free energy alternatives.

Fracking is a technique used to extract hydrocarbons trapped in certain kinds of rock. It has become widespread in the expansion of shale gas extraction. Shale gas is a natural gas trapped in impermeable shale rock, so simple drilling is not enough and the rock must be fractured to allow the gas to escape.

Hydraulic fracturing uses pressurised fluid to free trapped gas. Wells are drilled and the fracking fluid injected into them under high pressure to crack the rock. The fracking fluid consists of water, sand and chemicals. Millions of gallons of water are used to frack a well. Fracking isn't just used for extracting shale gas; it can be also used for shale oil and coal bed methane.

Many parts of the world, including the UK, lie above regions of shale, a layer of rock containing deposits of oil and gas thousands of feet beneath the surface. The fossil fuel industry has known about these resources for most of its history, going back to the time when the industry was young and vigorously testing the limits of what could be extracted from the earth and burnt for fuel. In fact, the mechanical principles of fracturing rock to release oil or gas were developed in Titusville, Pennsylvania at the birthplace of the US oil industry and patented in 1865-6. Fracturing using water, as is done today, dates back to the 1930s.

Hydraulic fracturing may not be a new, innovative technology, but neither is it a tried and tested, proven one, for the reason that it has never been carried out on the scale now proposed in the specific geological conditions of the densely populated UK. What has changed is that the development of horizontal drilling at a time of decline for conventional fossil fuels means that profits from shale can be made – if the scale of operations is large enough. In 2000, US shale gas fracking represented just 1.6% of natural gas consumption. By 2010, this had risen to a remarkable 23.1%, with devastating consequences for local eco-systems forcing entire communities to up sticks and move.[1]

In November 2012, the ConDem government announced that fracking to extract shale gas would resume, having been suspended since late 2011 following measurable earthquakes near Blackpool in Lancashire. These coincided with fracking by Cuadrilla in the local Fylde area and scientists believe there was a definite link.

To date this is the only attempted fracking operation in the UK, but many others are set to follow. In July 2013, Tory chancellor George Osborne announced substantial tax breaks for shale gas operators in Britain, reducing the rate on revenue to 30% compared to 62% charged to oil and gas drillers operating in the North Sea (and these rates are themselves already heavily subsidised).

Then in January 2014, prime minister David Cameron announced what is essentially a crude bribe to persuade local authorities to take up the fracking cause. He announced that councils will keep 100% of business rates they collect from shale gas sites – double the current 50% figure for other business. This amounts to direct funding by central government. This is one side of the government's close partnership with the industry which is also offering inducements. These include a £100,000 one-off payment to the community when a test well is fracked and a share of revenues if gas is extracted. Cameron also said that under new rules, energy firms could opt to make direct cash payments to local residents or into trust funds managed by communities. None of these sweeteners has stopped the momentum against fracking building in local communities.

In addition, the government has established an entire new department – the Office of Unconventional Oil and Gas – to handle the process of licensing proposed new extraction sites. This department will deal not only with fracking, but also with coal bed methane capture, which is proposed at Airth in Falkirk and elsewhere.

As a result, there has been a surge in planning applications by gas companies to start drilling in some of Britain's most treasured rural areas. Cuadrilla Resources has been trying to conduct exploratory drilling at Balcombe, West

Sussex, which generated a massive community resistance campaign centred around a camp pitched outside the drill site entrance, making national and even international news..

Meanwhile the UK Coal Authority has issued five licences to Cluff Natural Resources (CNR) for deep underground coal gasification (UCG). The process would re-open British coal-fields for exploitation, but without the need for miners. CNR is owned by Algy Cluff, who made his fortune in the 1970s from the North Sea oilfield, owned the Tory *Spectator* magazine, and was a director of the Thatcherite Centre for Policy Studies.

CNR has raised £2m for the five projects and is pursuing planning applications – in the Firth of Forth, in Scotland; the Loughor Estuary in Carmarthenshire, South Wales; the Dee Estuary, in North Wales and Merseyside; and at Whitehaven in Cumbria. The unique selling point of UCG is that it can tap hard to reach coal reserves which in Britain runs to billions of tonnes. The process involves deep drilling into coal seams and using horizontal drilling techniques similar to those involved in fracking.

Either air or pure oxygen is pumped down one well, and the coal is set on fire underground. The resulting gases are piped to the surface, where a mix of hydrogen, methane and carbon monoxide known as 'syngas', can be separated for burning in generators or for use in fertilisers.

UCG has been tested in the past on a scale significantly smaller than is proposed in Britain. These attempts have been plagued by disaster. Recent experiments in Australia resulted in two out of three plants being shut down in 2010. One, in Queensland, exploded after just five days. Carcinogens benzene and toluene were then found in ground water and the fat of grazing animals. Previous tests in the US and Europe have also been plagued by explosions and groundwater contamination.

A third process is coal bed methane capture and Dart Energy submitted a planning application to Falkirk and Stirling Councils for 14 well pads with 22 coal bed methane (CBM) wells, pipelines to connect the sites, a gas processing and water treatment facility and a waste outfall into the Firth of Forth. The site covers a large area between Larbert and Airth with drilling planned 20 metres from some homes and going underneath many more. Both councils rejected the application, but Dart Energy appealed to the Scottish government, and a public inquiry is planned.

CBM pumps out water, decreasing pressure in the coal seam, which releases methane to flow (it is hoped) up the well. The concern is that once released, the methane can go anywhere, emerge anywhere – into water supplies or even leaking from former mine shafts. The whole central belt of Scotland is littered with these. Dart claim that CBM is not the same as fracking, but the reality is that most CBM wells end up being fracked as as well, when the gas flow slows.

A report by David K. Smythe, Emeritus Professor of Geophysics at Glasgow University, says the Dart plan poses a threat to groundwater resources over the entire proposed development area, and "there is the additional risk that fugitive methane may even reach the surface", and, therefore, in his view, "the development should not be permitted". There is plenty of evidence that he is right. A massive CBM development in the Powder River Valley in Wyoming has polluted groundwater and rivers with toxic chemicals. In Queensland, Australia, the Tara field has over 5,000 wells, and a report from the Mines Department showed more than half of them are leaking.

The tactics of the police in response to protests reveal that the British state is working in league with the fracking companies, a close cooperation between industry and government that is rarely observed so openly on the front line. At Balcombe, the strong arm of the state stepped in to disrupt and break up the peaceful camp, arresting dozens including Green Party MP Caroline Lucas.

First hand reports describe officers pushing protesters along with undue force, and "snatching" protesters at night. Greater Manchester Police have carried out violent attacks on protestors at Barton Moss in Lancashire. Solicitor Simon Pook, who was acting as a legal observer at Barton Moss, wrote: "What I have witnessed today at Barton Moss has confirmed my greatest fear ... that Greater Manchester Police appear to have discarded the European Convention of Human Rights into the gutter, and replaced it with political policing, re-enforced with overt aggression... I am firmly of the view that what took place today appears to be political policing in favour of a corporate agenda."

At the same time as the rush to extract new fossil fuels takes off – not only in the US and UK but right across the world, from Algeria to Canada – the world faces an ecological crisis. The Intergovernmental Panel on Climate Change (IPCC) in its 5th report released in September 2013, issued a dire warning that climate change is accelerating now, heading fast towards catastrophic levels. They expressly state that global greenhouse gas emissions need to be cut at a rate of 10% per year, starting now, for a chance of the global temperature rise stopping at or close to 2ºC.

Nevertheless, governments all over the world continue to demand extraction of more oil, coal and gas. How and why do they legitimise policies to burn more fossil fuels? Why are renewables not considered a better course of action, both for the sake of the climate, and economically?

The simple reason fracking is attracting the support of governments worldwide is that it is seen as a quick fix solution to the problem of diminishing supplies of fuel to drive the capitalist system. Around the world, conventional oil and gas fields are running dry and are simply not supplying enough to fuel a global economy based on using ever-increasing amounts of energy to produce electricity and transport goods and people. In the UK, North Sea production peaked in 2003 and has been declining since. In this context, it becomes easy for a government to argue the need to exploit a new energy source, and all the better if that source is within its own borders.

In Britain, the ConDems are desperate to launch a strong shale gas industry, claiming that it would lead to an economic boost and bring down gas prices for people. This seems an attractive proposition if, for example, you are an older person, unemployed or in a low-paid job and cannot afford to heat your home because gas prices are so high.

The second reason relates to the use of gas to generate electricity. The UK uses large amounts of natural gas in power generation. The switch from coal to gas in power stations has been achieved through updating the generating technology over the last few decades and has been the main factor in the small reduction in the country's CO_2 emissions.

But it also means that electricity generation is now totally reliant upon either coal or gas – aside from the renewables sector, which remains small due to underinvestment – and nuclear power. The older nuclear power stations are set to close and so are many older coal-fired power stations. New nuclear and renewable energy cannot be activated quickly enough to make up a sudden shortfall in gas supply. For this reason, the government has seized on the prospect of shale gas and coal bed methane as a power source, amidst concerns that oil and gas supplies are running out.[2] Warnings are growing of an energy crunch. No government wants to be in power should that happen, and the ConDems are no exception.

However, whether swiftly drilled shale reserves could actually solve the problem of power cuts is questionable. Fracking is only the first part of the production process for shale gas. The infrastructure has not been built to pipe, compress and process what comes out of the wells in order to turn it into the final product. Shale gas will generally require more processing because it is a lower grade fuel, meaning more impurities and other obstacles that require chemical treatment. There is a limit to how quickly this can be achieved.

It seems that the gas industry itself doesn't expect the UK to be at a production stage until the end of the decade, generally assuming this in their projected figures. In addition, the full fracking job has been attempted already by Cuadrilla in Lancashire, where it was a disaster. The company has made enemies of the local community and damaged the well it drilled, stalling the prospect of its becoming Britain's first producing well.

This all raises the possibility that the dash for gas is more of a long-term strategy than a short-term one. After all, coal can always replace gas in the very short term. This happened in 2012, causing a 3.9% increase in CO_2 emissions for that year.[3] The government would attract criticism for going back to a dirtier fuel, knowing the impact it has on the climate, but would likely be able to defend the policy.

The assumption is made that shale gas as an electricity source would be a stop gap, filling in while a transition is made to renewables (and maybe nuclear). In reality, it is clear that the government views it as a major part of the energy future. They have commissioned 30 new gas-fired power stations, without waiting for scientists to pass judgement on whether this is a good idea, and whether these have any place in an energy strategy aimed at delivering deep cuts in emissions.

Nor do they provide a "transitional fuel" – a cleaner source to power the economy while the transition is made to a low-carbon future. This is the claim made by many industry figures, most notably Lord Browne, the Cuadrilla part owner and former BP chief executive. While the argument may be valid for conventional gas, shale gas and coal-bed methane are likely to increase global emissions disastrously. In brief, it is because the climate change impact of fracking is not only about the CO_2 produced in the burning of the gas, but much more about the associated fugitive emissions of methane from the drilling and production process.

Experts have blown apart claims by prime minister David Cameron that shale gas exploitation will cut fuel bills and create thousands of jobs. Bloomberg New Energy Finance (BNEF) submitted evidence to MPs showing that fracking will not bring down gas and electricity bills as it is likely to cost between 50-100% more to extract than in the US.[4]

"The reliance on continued imports will ensure that UK gas prices remain tied to European and world markets and so the direct impact of shale on the cost of electricity in the UK will be limited," the report says. In August 2013, Cameron told the the *Telegraph*: "Fracking has the real potential to drive energy bills down ... many people are struggling with the cost of living today. Where we can relieve the pressure , we must. It's simple – gas and electric bills can go down when our home-grown energy supply goes up."

A month later, the justice secretary Chris Grayling, asked to name one policy the Tories would propose to help people with the cost of living at the next election, said: "Good question. I think my answer to that would be the rapid acceleration of shale gas. That could make a huge difference. The sooner we accept it is a powerful source of energy for the future of this country that could see us through, the better."

But they were making it up as they went along, a point reinforced by the statement by Ed Davey, the Lib Dem energy secretary, who warned shale will not have "any effect" on energy bills given Britain is part of a global energy market. "North Sea gas didn't significantly move UK prices – so we can't expect UK shale production alone to have any effect," Davey said. Then respected economist Lord Stern, author of the ground-breaking 2006 report on climate change, dismissed Cameron's claims. In an interview with the *Independent*, he said: "I do think it's a bit odd to say you know that it will bring the price of gas down. That doesn't look like sound economics to me. It's baseless economics." In February 2014, Osborne himself played down expectations that household gas bills would fall when he gave evidence to a parliamentary committee. He acknowledged that "we more closely track the worldwide gas price than the US does and we have less ability to detach ourselves from the worldwide gas price".

As to the government's original claim that Britain's shale gas could create 74,000 jobs, this forecast was cut by two-thirds by AMEC, the engineering consultancy that is advising DECC, the energy department. AMEC predicted at a meeting in Whitehall that 15,900 to 24,300 full-time equivalent jobs – direct and indirect – would be created at "peak construction" by the shale gas industry. The very same figures were quoted in a government press release published in December 2013. But this did not stop Cameron from repeating the larger figure when he visited a drill site in Lincolnshire early in 2014. Whitehall referred you to the crazy free-marketeers Institute of Directors' website to back Cameron's false claims, thus proving once and for all that he is the snake oil salesman for the fracking companies.

It might seem reasonable to believe the UK could exploit large shale reserves – after all, this is what has happened in the US. But in the frenzy to extract, supporters of fracking have forgotten that shale gas is a classic bubble. There is growing evidence that the US shale bubble is set to burst, even as energy analysts continue to forecast growth over the coming decades. US geoscientist David Hughes recently warned that the cheap price bubble will burst within two to four years. He notes that the US Department of Energy has recently downgraded its estimate of the size of shale reserves.[5] Where before land prices above shale were soaring, now drillers are pulling back from buying any more and prices have collapsed.

Back in the UK, the British Geological Survey (BGS) reported in June this year that there was enough shale gas under Britain for 40 years of consumption at the present rate. If the US boom could go bust after just 10 years, this projection would seem wildly optimistic. In fact, it is based on misrepresentation of the BGS figures. The figure of 1,300 trillion cubic feet (Tcf) in the central UK is an estimate for the total resource, not the reserves that can be extracted. "Resource" in this case refers to all shale regions containing gas. As the Oil Drum website stated:

> "The BGS report, unfortunately, only addresses gas-in-place (total resources) and not extractable resources (technically recoverable resources) much less reserves (commercial supply). The most-likely reserve potential of the Bowland Shale is only about 42 Tcf (3% of gas-in-place) after applying methods used by the US."[6]

Even this appears to be an optimistic view. Another analysis cites a figure of 25 Tcf as the technically recoverable supply. Of course, these studies are nowhere to be found in mainstream media reports even though they come straight from the industry itself.

So even if the UK fully developed its shale gas industry, the result would be 7-13 years of gas. There is no guarantee that this would even go to UK consumers, because the gas would be sold on the global market to whoever pays the wholesale price. But even if all the gas was used here, Britain would still be dependent on imports of gas in the long run.

Unconventional gas is a doomed energy strategy – and the worst possible response to climate change. In essence, this is because huge inputs of energy are required at every stage of the capitalist production process for all basic commodities. For decades, this energy source has been fossil fuels. With globalisation it has got to the point where CO_2 concentrations in the atmosphere have increased from 350ppm to 400ppm in just 25 years.

World grain production, dominated by global corporations who control farming with large scale, inefficient industrial methods, depends on petroleum-based fertilisers. All world trade depends on the movement of people and goods by air, road and sea. And above all, electricity for houses and industry is still being generated by burning coal, oil and gas. Now, for the first time, the supply of energy is running out. Apart from the role of fuel in production of goods and in agriculture, the capitalist economy operates on the principle of investing in energy itself, in order to extract profit. This means the energy gain must be greater than the input of energy used in the extraction process.

As the cost of energy rises, it becomes profitable to drill unconventional fuels like shale gas which were previously unprofitable. But with these "new" products, the expected energy gain is much smaller because unlike traditional oil and gas, they are not only hard to access and retrieve, but also to process in order to become saleable natural gas. Author David Hughes estimates that the Energy Return on Energy Invested (EROEI) for shale gas is likely to be 5:1 or less, compared to 7.6:1 for natural gas.[7]

The remaining impetus for fracking is provided by global finance and the big banks. Their short-term strategies, like those of other powerful transnational corporations, prioritise profit even if based on unsustainable boom and bust. While fracking does not deliver an energy profit, and therefore no advantage to the capitalist economy in absolute, physical terms, it can deliver a profit through financial speculation.

The insatiable global demand for energy drives speculation in shares of the fracking companies, causing an asset price bubble that, like all bubbles, will burst. When over-optimistic production goals are not met, and it becomes impossible to go on producing the gas at the prevailing low market price, the massive debts of the fracking companies will become another phase of sub-prime junk debt.

The shrewdest investors will sell before this happens, resulting in a total price collapse. This is already underway in the United States. A fracking-related land gamble, where speculators were buying up land hoping that at some point the fracking companies would pay to move in, has already collapsed. In effect, the fracking companies are almost purely financial vehicles for profit, controlled remotely by a secret few who are poised to take any benefit.

Despite claims by the Office of Unconventional Oil and Gas that they can ensure "safe and sustainable" production of Britain's shale gas reserves, the secret of profit for the industry is exploiting the input cost/output price gap, and that will not lead to safe working or a stable energy supply.

2
Planning system silences the voice of the community

When it comes to the planning process, the system is weighted heavily in favour of fracking companies. Local authorities have little room for manoeuvre in the wake of current law and government "guidelines".

For example, the guidelines state: "Mineral extraction is essential to local and national economies. As stated in paragraph 144 of the National Planning Policy Framework, minerals planning authorities should give *great weight to the benefits of minerals extraction, including to the economy, when determining planning applications.*" [emphasis added]

Originally, the government had intended to take shale gas planning decisions out of the hands of local authorities. After ministers announced in late 2012 that the dash for shale gas was on, they wanted to amend the Growth and Infrastructure Bill then going through parliament. Their aim was to designate shale gas decisions a matter of national energy policy, placing them firmly in Whitehall's hands.

But when the legislation – which deregulated a whole raft of planning restrictions in relation to housing and other developments – went through, the government appeared to back down and withdrew the proposal. This crafty move aimed to show that the government believed in "localism". The sting was in the tail, however. Ministers made it clear that the new legislation gave them powers to include shale exploration decisions should they want to do so at a later date. A statement insisted: "Shale gas extraction has yet to take place at a commercial scale in this country and, as it develops, the government will ensure that an effective planning system is in place."

James Taylor from Simmons & Simmons, a global law firm that numbers energy firms amongst its clients, believes it would be relatively easy to bring shale gas into the fast-track system at a future date. "The government will wait and see what happens with initial shale gas developments," he told *Planning* magazine, "but this remains in their pocket to pull out if local planning is a significant barrier to shale gas."

As applications for test drilling spread around the country, planning departments are on the back foot. Any company wanting to conduct exploratory drilling in the UK has first to obtain a licence from the Department for Energy and Climate Change. It also needs a variety of permits from the Health and Safety Executive and the Environment Agency, and will also have to apply for planning permission.

Planners will have to consider local issues, such as impacts on landscape and traffic, as well as whether there is a need for an environmental impact assessment (EIA). If the company decides to go ahead with commercial operations, it will need to submit further planning applications to allow permanent operation. Already depleted of staff as a result of spending cuts, many departments simply do not have the same expertise that is at the disposal of the frackers. For example, when it came to EIAs, Lancashire County Council actually asked Cuadrilla itself to perform them on all their planned test sites and, naturally, the company agreed. Cuadrilla appointed consultants Arup to carry out the work.

As bids for test-drilling or commercial operations are mineral applications, they will be considered by county councils, which become Mineral Planning Authorities for the purpose. County councils are the most remote tier of local government, with offices often located in small market towns, making them harder for campaigners to access than their local or parish council.

After the new planning laws went through, in July 2013 the Department for Communities and Local Government (DCLG) stepped up the pressure. It published "technical guidance" for planning authorities on deciding applications for the exploration and extraction of shale gas. And planning experts agree that the so-called guidance essentially amounted to a green light for the companies.

The guidance covers whether operators need to provide Environmental Impact Assessments, how planning will interact with other bodies involved with the regulatory regime covering shale gas, including the Environment Agency and the Health and Safety Executive, and site restoration. At first glance, it seems harmless enough. The guidance "is not intended to replace the need for judgement by minerals planning authorities and those making planning applications". Nor is it intended to be a source of definitive legal advice. But as you plough through, the message is clear enough. The "guidance", which becomes part of the planning process and has some force in law, insists:

- planning applications for exploration of shale gas should not take account of "hypothetical future activities" – actual extraction
- planning authorities should not consider demand for, or alternatives to, oil and gas when determining applications
- councils should understand that government policy is clear that energy supplies should come from a variety of sources
- planning authorities should give great weight to the "benefits of minerals extraction, including to the economy".

And just to emphasise what the priorities are, Baroness Hanham, a DCLG minister, said: "Effective exploration and testing of the UK's unconventional gas resources is ... key for understanding the potential of this industry. The government is *creating the right framework to accelerate shale gas development* in a responsible and sustainable way." [emphasis added]

There was outrage in one quarter, at least, when the guidance appeared just before parliament broke for a long summer recess. Green MP Caroline Lucas condemned the lack of consultation and told a special debate on shale gas that it was "pretty appalling" that the guidelines were set to come into force without consultation, "denying communities that stand to be affected by fracking any say in the new process":

> "It is clear that ministers and the fracking firms, which are, sadly, increasingly indistinguishable, are keen to press on rapidly, but it is wrong to refuse to consult on new planning guidance aimed at making it easier for developers to cast aside community concerns. Even from a perspective of due procedure, I cannot see how the decision to deny communities a say in their new planning rules is remotely in line with the government's own definition of circumstances in which consultation is unnecessary."

Lucas explained that issuing guidelines without consultation was normal only if they were minor or technical amendments to regulations or existing policy frameworks, or where consultation had taken place at an earlier stage. "Many of my constituents have e-mailed me over the past few weeks to call for a full public consultation, as well as for new planning rules that are strong on tackling climate change and follow the precautionary principle when it comes to issues such as groundwater contamination," she added.

Her constituents' voices, along with those around the country threatened by fracking, have effectively been silenced by the government.

A report in *Planning* shortly after the guidance came out reported that professionals were concerned about "bias" and expressed fears that the planning system was, as a result, weighted in favour of granting permission. James Taylor said: "The way the document reads, it could be seized upon by anti-frackers as unduly biased. It pulls together pre-existing advice but puts the most positive spin possible."

For example, he said, the guidance states that Environmental Impact Assessments (EIAs) are unlikely to be required for exploratory drilling. "This guidance is almost predetermining local authorities not to require EIAs," Taylor said.

Don Gateley, principal planner at property firm Savills, said the guidance was akin to a presumption in favour of shale gas. "Rather than just introducing controls over how decisions would be made, the guidance implies that government wants to see them go through," he said.

There are up to seven other bodies that could be involved in shale gas development, and mineral planning authorities (MPAs) should assume that regimes by other bodies in the consenting process will operate effectively, the guidance states. Planners should not carry out their own assessment of issues such as earthquake risk, well design and construction and chemical content of hydraulic fracturing fluid. They should instead take advice from the relevant regulatory body, the document adds. Planners will be on "thin ice" if they refuse applications on these issues and will end up at appeal, Gateley said.

Friends of the Earth criticised the guidance for "cherry picking" from the National Planning Policy Framework (NPPF) but ignoring climate change. "It's contradictory and confusing – the NPPF is clear that you have to consider climate change," said Naomi-Luhde Thompson.

As the hot summer rolled on into September, the government stepped in again with plans to change the current notification requirements on underground gas and oil applications to make it easier for the corporations, naturally. At present, companies have to serve a notice on any other person who owns or is a tenant of land to which the application relates. They also have to publish a notice in at least one local newspaper and display a notice in every parish affected ahead of submission.

But under the proposals, the requirement to serve notice on individual owners and tenants of land would be dropped where solely underground operations would be involved. So, in practice, companies would be able to apply for planning permission to drill for oil or gas under people's property without first notifying them!

The document says the government "considers that underground operations for the winning and working of oil and gas are different in character from other existing forms of development" as such operations are at "significant depth below the surface". It also states that, at the time of an application, it can be difficult for the developer to precisely define the underground area of land where drilling will take place and the area of land from which the oil and gas will be removed.

"This means that, because of the uncertain route of the underground works, the area of land identified in a planning application which is to be developed has to be drawn widely to ensure that it is broad enough to cover any potential route of lateral drilling and area of working. A widely drawn area would necessarily require the notification of significant numbers of owners," the document says. In light of this, the document says, the government "considers that it is unreasonable and impractical to require applicants for planning permission for underground oil and gas working to serve notice on owners and tenants of land across such a widely drawn area".

A Lords' committee found that the government had rushed through the changes without proper consultation. Planning ministers laid the "order" to introduce the changes before Parliament on December 20, while MPs were away on Christmas holiday, and brought it into force on January 13, a week after they came back, giving "scant opportunity" for scrutiny. Public consultation on the changes went on for just six weeks – half the time that the government's own guidelines suggest would have been appropriate for a "new and contentious" policy such as procedures related to fracking, the Lords secondary legislation scrutiny committee said.

"Given that 'fracking' is a highly controversial technique, and that the order streamlines procedures for notifying interested parties whose land may be affected by the technique, we find it regrettable that the opportunity for Parliamentary scrutiny was curtailed in this way," the committee said. Ministers, naturally, failed to disclose to Parliament the scale of opposition to the changes in the public consultation, which saw seven responses in favour and 155 against.

Ministers are now also planning to change the trespass laws to make it easier for fracking operations to take place without homeowners' consent. The planning system is now so heavily slanted in favour of fracking that any talk of localism or respecting communities is just that – talk. In practice, the state at national level has instructed the local state to get on with smoothing the path for the frackers. Democracy this isn't.

CASE STUDY: The Balcombe experience

A letter to the Editor (*Guardian*)[8] from Professor Lawrence Dunne, Professor Alan Rew, Jackie Emery and nine others: "Tim Stone's perception of UK regulation of shale and oil and gas extraction is very much at odds with our experience here in Balcombe. The planning permission allowed 60 passages per week of drilling trucks through our little village. When Cuadrilla pleaded that they had meant 120 (in and out), the regulators just rolled over. The drill is a few hundred metres from houses and the regulators have known for weeks that the total noise at night exceeded that granted in planning. Despite multiple complaints and requests, they made the figures public only recently, and then thanks to pressure from the community. It was Friends of the Earth who forced the Environment Agency to consider the need for mining and radioactive waste licences, and pointed out the ambiguous legal position of the horizontal well. As a result, Cuadrilla must now reapply for full planning permission to test its well. The Environment Agency has issued flaring permits for emission of air pollutants over the populace of Balcombe, but without publishing emissions limits. Of equal concern are Defra's published proposals for significant reduction of local air quality monitoring in Britain. This conveniently fits with the coalition's more-fracking-less-regulation agenda."

3
Resistance movement threatens the frackers

Communities are uniting across the world in actions aimed at thwarting the shale gas corporations and exposing the anti-environmental consequences of their operations. October 19 was Global Frackdown Day and in 2013, activists from 26 countries took part in around 250 protests in cities across the world to demonstrate against the destructive impact of fracking.

Around 1,000 people in Pungesti, Romania, protested against US oil giant Chevron's plan to start drilling outside their village. The energy firm had received permission from the local authority but protesters shouted "Chevron, go home" and "We say no to shale gas".

"I am against shale gas exploitation because of the chemicals used in fracking," said Vasile Ciobanu, 25, who has returned to Pungesti after working abroad for three years and now lives a few hundred metres away from the proposed well site. "I don't think the company and Romanian officials are thinking about what could happen to people who live here." Chevron temporarily suspended operations in the wake of the protests.

Dozens of demonstrations were also held across the US. Protesters took to the streets near Raritan River in New Jersey, while others gathered in New York against a massive fracking programme planned for the state. California activists marched and participated in a bike event to protest against fracking near Ballona Creek. A demonstration was also held in the city of Oakland.

Around 250 demonstrators gathered in three Canadian cities in New Brunswick. Some 100 protesters blocked one lane of the highway near the town of Rexton.

In the days leading up to October 19, the Royal Canadian Mounted Police violently removed barriers in the town erected by members of the Elsipogtog First Nation tribe, arresting at least 40 people. The Elsipogtog say that fracking could irreparably damage their land and the surrounding area.

First Nation leaders said police provoked the confrontation in which several cars were burned, by arriving with guns drawn. "It is really very volatile. It's a head-to-head between the people and the RCMP right now and the warriors are in the middle surrounded by the RCMP and then the RCMP are surrounded by the people," said Susan Levi-Peters, a former Elsipogtog First Nation chief.

The Quebec anti-fracking movement is exceptionally well organised and co-ordinated. In a guide to the campaign, *Civil resistance as deterrent to fracking*, organiser Phillippe Duhamel writes:

> "It's win before you fight. Using an innovatively designed civil resistance campaign as a nonviolent deterrent, the people of Quebec have so far been successful in defending their land against hydraulic fracturing. Over the course of three years, plans to drill some 20,000 shale gas wells along the St. Lawrence River, between Montreal and Quebec City, have been thwarted to the point of being recognized as a de facto moratorium on this form of extreme energy extraction."[9]

La Campagne Moratoire d'une Generation (MDG), the One-Generation Moratorium Campaign, was founded on a mission to prevent dirty energy from being developed in Quebec. One of the highlights was the month-long walk organised in the spring of 2011 along an itinerary closely following the areas claimed for fracking by the industry.

While the marchers were on the road, a Bill was presented and adopted for a reduced five-year moratorium on shale gas development under the St Lawrence. On the eve of the walk entering the city, the environment minister declared for the first time that no more fracking would be allowed in the province at all, until further notice.

> "More than two years later, the de facto moratorium is still standing. We call it a citizen moratorium, because it was clearly the result of grassroots organising and popular mobilisation."

The fracking industry is clearly concerned about the impact of media-savvy protests on local communities because they do not want to run the gauntlet of ordinary people on a day-to-day basis. So London-based Control Risks, a global risk assessment consultancy for industries and governments, was commissioned to carry out an in-depth study of anti-fracking groups around the world entitled, The Global Anti-Fracking Movement: *What it Wants, How it Operates and What's Next.* In a frank assessment, the consultants admit:

> "Direct action serves both strategic and tactical purposes. Strategically, it attracts media attention, raising public awareness of hydraulic fracturing, and thereby increasing receptiveness to anti-fracking messaging and aiding activist recruitment. Demonstrations, days of action and non-violent civil disobedience provide impetus and focus to the anti-fracking movement, helping to mobilise grassroots support, and generating solidarity both locally and globally."

Activists joined protests in Spain, France, and Ireland. At least 13 separate demonstrations were held in UK cities. Thousands of people took part in protests in Sussex over the summer of 2013, helping to turn public support in the UK against shale gas for the first time. Polling by the University of Nottingham had shown support for shale gas extraction in the UK steadily rising for more than a year, peaking at 61% in favour in July. But that number fell in September to 55% and has continued to fall.

Professor Matthew Humphrey, who worked on the Nottingham study, said high-profile protests could hit public support for fracking: "This may have important implications for the politics of fracking in the UK, if the anti-fracking lobby come to believe that highly visible forms of protest at potential sites for hydraulic fracturing are the most effective means of changing the public mood."

Even the government's own figures have confirmed the public's support for renewable energy is increasing while those in favour of fracking has slumped to a record low. The DECC Tracker of Public Attitudes, released in February 2014, reveals the percentage of surveyed households that were in support of renewable energy, such as wind and solar, has now increased to 77%. However, despite the government's public relations campaign on behalf of the fracking industry, only 27% said they were in support of shale gas exploration.

But it is not only in Britain that the worm is turning. A major survey of nearly 23,000 individuals, businesses and public authorities by the European Commission in September found that 40% were against developing shale gas as an energy source in Europe. Only 33% were in favour regardless of environmental safeguards. The poll was probably a factor behind the narrow vote in the European Parliament to strengthen fracking regulations, contrary to the plans of the European Commission.

Legal challenges are mounting too: Greenpeace has launched a website ironically named wrongmove.com where you can find out if your neighbourhood is a fracking target. Landowners can also sign up to join an action to block moves to drill under their land without permission. Greenpeace is calling on local residents who oppose shale gas or oil exploitation near them to join together to assert, what they claim, are common law rights to forbid the "trespass" of energy firms from conducting activities below their property.

The legal challenge, which was launched in Preston, was joined by residents from the Fylde and campaigners from Balcombe. Lawyers for Greenpeace claim that a Supreme Court ruling in 2010 in which Mohammed Al-Fayed sought a share of the proceeds from an energy firm after it drilled for oil under his Surrey estate paves the way for mass action by concerned citizens. Five justices rejected the former Harrods owners' demands for compensation and damages but unanimously upheld the claim of trespass.

Andrew Pemberton, a Lancashire dairy farmer said he was worried for the ecology of the Fylde near Blackpool and would be joining the protest. "I'm supplying milk to 3,000 households, and if for any reason my water became contaminated, my business would be ruined and my livelihood destroyed, as well as the livelihoods of the 16 families who work for me. Fracking is dangerous and short-sighted. We should be keeping this gas in the ground."

Concerned Communities of Falkirk raised an incredible £50k from thousands of supporters right across the country giving small amounts so that they could have a top QC to represent them at the inquiry into Dart's application to launch a coal bed methane capture operation in their community.

Landowners in the Sussex Downs National Park are mounting a "legal blockade" to block a potential fracking site. Solicitors for residents near Fernhurst, in West Sussex, have written to Celtique Energie and the energy secretary to explicitly deny permission to drill under their land. Marcus Adams, one of the landowners involved in the legal blockade, said:

> "People right across the country have legitimate concerns about the impact of fracking on their communities – from water contamination to air and noise pollution from heavy lorry traffic – but that all this is happening in a National Park just doesn't bear thinking about."

4
Trapped by a corporate state web

A web of connections stretching from corporations, to lobbyists, through to ministers and to departments of state ensure that the business of government is business itself. Britain is not so much a democracy as a corporatocracy, where the interests of companies are deemed synonymous with those of the state.

Money does not have to change hands to buy influence. Business is kicking at an open door and has done so for a long time. What has changed, however, is that the state has abandoned any sense of neutrality between citizen and corporation. Profit-driven globalisation has spawned a fully-fledged market state in place of the old welfare state.

Throughout 2013, the government and the fracking industry worked seamlessly to try and overcome public opposition. Emails obtained by Greenpeace showed how Whitehall officials worked with the companies to manage the public's hostility to fracking. In one email, the Department of Energy and Climate Change (DECC) apologises to the UK Onshore Operatives Group (UKOOG): "Sorry to raise your blood pressure on this subject again, no expletives please!" Another email from Centrica to DECC officials warned that officials in Lancashire County were not convinced there was enough government regulation for shale gas. Of course, officials described the email exchanges as "right and proper" – which they are if you believe the business of the state is business itself!

Caroline Lucas has described what she calls "inappropriate corporate influence in government" that is "at the core of the coalition's irrational enthusiasm for shale gas and fossil fuels more widely". But the state considers the relationship with the industry entirely appropriate and rational. Speaking in a parliamentary debate on fracking in July 2013, Lucas told MPs:

> "Worse still, it means that we are seeing policies designed to maintain the status quo, where power is literally and metaphorically concentrated in the board rooms of big energy companies such as the owner of British Gas, Centrica, which recently bought shares in Cuadrilla."

On this point, Lucas is entirely right! Lord Browne, the former BP boss, is chairman of Cuadrilla Resources. At the same time, Browne holds an unpaid position in the Cabinet Office, where he is apparently responsible for convening meetings of senior business figures. Browne reports to the minister for the Cabinet Office, Francis Maude, in whose Sussex constituency Cuadrilla plans to drill. In August 2013, *The Independent* published an investigation by James Cusick into how the corporations influence policy.[10] He explained how the government came to embrace fracking as the best thing since sliced bread. At the end of 2011, Browne arranged a series of meetings about fracking with senior ministers in DECC. Arrangements were also made for the company's senior management to meet DECC's minister in the Lords, Lord Marland and another minister, Charles Hendry. Cusick says:

> "Soon, the company went for another tack. Finding their work with the energy department wasn't working, they went for the Treasury instead. And though they are understood to not have been involved in face-to-face DECC meetings, corporate affairs consultants – lobbyists working for the drilling company – played their part behind the scenes.
> With their help, academics were brought in to analyse and publish data on the scale of shale gas's UK worth. Trusted journalists were contacted. Lists of those with opinions that mattered, including MPs, were drawn up. A lobbying strategy, aimed at changing and utilising the opinion of key players, was developed. The focus, with the Treasury at the centre, was on reducing the UK's energy bill, ending the reliance on gas from abroad. In the wake of these meetings, government assistance in ensuring Cuadrilla maximised the financial benefits of the energy 'fracking' licences the company holds at 10 sites throughout the UK was said to be critical in convincing the Treasury that fracking could solve Britain's energy problems."

Thousands of local people and supporters from across the nation, turned out in August 2013, to protest against Cuadrilla's test fracking site in Balcombe, West Sussex

They were met with a massive police presence, and legal threats from their elected representatives on the county council. Photographs Peter Arkell.

We now know the result of this enterprise. The government lifted the moratorium on fracking in November 2012 and new licences were approved. In 2013, planning laws were reinterpreted to smooth the path for fracking applications. In July 2013, chancellor George Osborne announced massive tax breaks for fracking corporations and senior Tories, from the prime minister down, began to sing about the wonders of shale gas. Job done.

All Browne had to endure was a protest outside his Chelsea home. Activists dressed in orange boiler suits and gas masks from anti-fracking group, Frack Off London, erected a 20ft drilling rig outside his house. Campaigner Adella Mason said: "This corporate-state love-in is going to make a few people very rich and leave the rest of us with thousands of fracking wells."

The love-in she refers to is about to get a whole lot worse under the proposed Transatlantic Free Trade Agreement (TAFTA) between the US and the European Union. The explicit aim is to create the world's largest free trade area, protect investment and remove "unnecessary regulatory barriers". Corporate interests are driving the agenda, with state bureaucrats doing their bidding on both sides of the Atlantic.

Analysing the plans for *Counter Currents,* Colin Todhunter warns: "There is growing concern that the negotiations could result in the opening of the floodgates for GMOs (genetically modified organisms) and shale gas (fracking) in Europe, the threatening of digital and labour rights or the empowering of corporations to legally challenge a wide range of regulations which they dislike. One of the key aspects of the negotiations is that both the EU and US should recognise their respective rules and regulations, which in practice could reduce regulation to the lowest common denominator."[11]

A report published by the Seattle to Brussels Network (S2B), exposes the trade deal. "Big business lobbies on both sides of the Atlantic view the secretive trade negotiations as a weapon for getting rid of policies aimed at protecting European and US consumers, workers and our planet," said Kim Bizzarri, author of *A Brave New Transatlantic Partnership*.[12]

"If their corporate wish-list is implemented, it will concentrate even more economic and political power within the hands of a small elite, leaving all of us without protection from corporate wrongdoings." The report also warns that the agreement could open the floodgate to multi-million euro lawsuits from corporations who can challenge policies at international tribunals if they interfere with their profits.

Where big business is involved, high-powered corporate lawyers are sure to be in the next room, or on the next plane to Europe in this case. In October 2013, the *New York Times* wrote a piece about a potential "lobbying bonanza" as a major American legal firm brought its European staff together near the EU's headquarters.

"Gathered at the Brussels office of Covington & Burling, a prominent Washington-based firm, were some of its lawyers and lobbyists, along with executives from some of the world's largest oil companies, including Chevron and Statoil. Their aim was to help shape the European Union's policies on the gas and oil drilling technology known as hydraulic fracturing, or fracking."[13]

They were meeting with Kurt Vandenberghe, then a top environmental official for Europe and a key player in the debate over fracking. The host that day was former Belgian diplomat Jean De Ruyt, now an adviser at Covington. "He and others on the recently expanded lobbying team there have delivered at least four senior European Union policy makers to the firm's doorstep in recent months, including a top energy official, who arrived in September with a copy of a draft fracking plan that has yet to be made public," says the report. "It's key to us to be ahead of when the political debate starts," Mr. De Ruyt said in an interview later. "Because by then, we can't have an impact." EU officials turning up with a draft fracking plan that no one else has seen – how blatant!

Because they are a law firm, Covington hides behind the cloak of client confidentiality and will not reveal who they are working for. But the firm is reported to be organising an industry group to offer EU officials "suggestions in drafting the rules", according to the *New York Times*. "There have already been preliminary moves by the Parliament suggesting it will demand strict oversight of the industry, an effort the lobbyists will try to derail."

The lobbyists were 100% successful. With the UK government applying considerable pressure, the EU abandoned attempts to regulate fracking. Plans for a new legally-binding directive were replaced with a set of non-binding "recommendations" covering protection against water contamination and potential earthquakes. Under the announcement published on January 22, 2014, every EU nation will be asked to produce a public "scorecard" within six months stating which recommendations have been implemented. The UK was joined by Poland in opposition to binding regulations.

In December 2013, prime minister Cameron wrote to the president of the EC, José Manuel Barroso, stating: "It is essential the EU minimise the regulatory burdens and costs on industry and domestic bill payers by not creating uncertainty or introducing new legislation." He added: "The [shale gas] industry in the UK has told us that new EU legislation would delay imminent investment."

In a leaked letter seen by the *Guardian*, the UK's top civil servant in Brussels. Ivan Rogers, a former banker at Barclays Capital and Citigroup, wrote in November that "seeing off" the proposals for new laws would require "continued lobbying at official and ministerial level using the recently agreed core script". Rogers advised that the government draw up a longer-term strategy to "manage the risks" when faced with a new European parliament and commission.

Keith Taylor, the Green party MEP for the south-east – including Balcombe in Sussex where major anti-fracking protests took place in 2013 – said: "It's deeply disappointing that the EC is set to publish proposals that will do nothing to protect EU citizens from the dangers of fracking. The UK government may be pleased with this result but those living near shale gas reserves will be very worried." Naturally, the industry was well pleased with itself. Ken Cronin, chief executive of UKOOG, the UK's onshore oil and gas trade body, cautiously welcomed the UK government's defeat of the EC proposal: "There are already nine different European directives covering the onshore oil and gas industry, so it appears to be good news that we don't have to comply with a tenth. However, we would need to see the detail before we can wholeheartedly welcome them."

Meanwhile, heading to Britain to drive the fracking industry forward is none other than Halliburton, the US oil services corporation that was central to the Deepwater Horizon disaster in the Gulf of Mexico. The energy company, previously run by former US vice-president Dick Cheney, has held secret talks with Celtique Energie, a British-based company planning to drill for oil and gas in the Sussex countryside, including on land that is within the South Downs National Park.

The potential involvement of one of the world's biggest companies in Britain's fracking industry indicates the accelerating scale of drilling envisaged across the country. Tony Bosworth, energy campaigner at Friends of the Earth, said: "People in Sussex have seen the impact of fracking in the US and the news that one of America's biggest fracking companies could be coming to the area will make them even more ready to fight drilling plans."

But it won't just be Halliburton they'll be up against. All the resources of the state are at the disposal of the fracking corporations.

5
The road to ecocidal suicide

Contrary to government and industry rhetoric, shale gas will actually increase global greenhouse gas emissions. This can only intensify the climate change crisis that the Intergovernmental Panel on Climate Change described in great detail in its 5th assessment published in September 2013.

Some of the key findings were that the planet is warming at a rapid pace without any doubt, that humans are causing it with 95% certainty and the last three decades have been the hottest on record. Since the 4th IPCC report (2008) an additional 200 billion tonnes of CO_2 have been released into the atmosphere and emissions are 60% higher than at the time of the first report in 1990. The reality is that current energy production and land use changes have already locked us in to going over the 2ºC warmer "red line".

Current emissions trends would take us over 4ºC global warming by the end of the century or sooner; at that point there is no potential for mitigation. Trends suggest there is an imminent risk of feedback events. For example, deep sea temperatures have been absorbing huge quantities of CO_2 and their ability to absorb more is dwindling, according to the National Oceanic and Atmospheric Administration (NOAA).

The Greenland and Antarctic ice sheets have steadily melted in two decades and glaciers are shrinking worldwide. Extreme weather is now commonplace, as the floods of 2013/2014 in England have demonstrated. So wilfully adding to emissions through the plundering of even harder-to-reach fossil fuels amounts to what some have dubbed "ecocidal suicide".

Supporters of fracking have argued that gas could replace coal as the primary fuel for power generation and cause some reduction in emissions. They based this assertion on the fact that gas produces less CO_2 per unit of useful energy generated than coal. But even if the UK could produce enough shale gas to substitute it for coal, this would not necessarily lead to a global reduction in emissions, because we are in a global market.

When US generating companies switched to cheap shale gas, the coal being produced in ever greater amounts thanks to the Obama administration's granting many new licences, poured on to the world market at cheap prices. In 2012 113 million tons were exported, the highest level in over 20 years, and nearly 10% of total US production. Asia's share increased from 2% in 2007 to 25% in 2012.

And the result? US emissions have decreased slightly from a 2005 peak but globally, emissions have soared in the same period. CO_2 concentrations in the atmosphere have risen from 380ppm in 2006 to 400ppm in 2013. Global consumption of coal increased by 25% between 2005 and 2011. An industry report predicts that world coal-fired power plant capacity in 2020 will be a third higher than in 2010.

In a report on US shale gas, Kevin Anderson of the Tyndall Centre for the study of climate change, concludes:

> "Given the global market for fossil fuels is growing and that global economic growth remains dependent on access to such fossil fuels, extraction of a new fossil fuel source is likely to depress overall fossil fuel prices and by definition increase demand i.e. catalyse an increase in absolute emissions. In this regard, and in the absence of meaningful emission caps, shale gas extraction within a market-based energy system will lead to an absolute increase in emissions."

A report for the EU written by UK firm AEA Technology plc and published in September 2012 concluded that the majority view was that emissions from shale gas are higher than conventional gas, but lower than coal. Electricity generated from shale gas produces greenhouse gases around 4% to 8% higher than conventional piped gas extracted within Europe.

Oil and gas wells leak and shale gas wells are no exception but with even worse consequences for the ecosystem. Because what they leak is methane, a greenhouse gas up to 72 times more potent than carbon dioxide. Figures from the US government and industry indicate that at least a third more methane leaks from shale gas extraction than from conventional wells – and perhaps more than twice as much.

"Compared to coal, the footprint of shale gas is at least 20% greater and perhaps more than twice as great on the 20-year horizon, and is comparable over 100 years," researchers wrote in the journal *Climatic Change*. Robert Howarth from Cornell University told BBC News: "We have used the best available data [and] the conclusion is that shale gas may indeed be quite damaging to global warming, quite likely as bad or worse than coal." In August 2013, a National Oceanic and Atmospheric Administration-led study measured a stunning 6% to 12% methane leakage over one of the US's largest gas fields, the Uintah Basin, which produces about 1% of US natural gas. In addition, there are large emissions from the industrial processes and activity that accompany shale gas drilling, on top of the emissions from burning the gas produced as the end product. A US Department of Energy analysis estimates that over its lifetime, a single well would produce around 5 billion cubic feet (Bcf).

To extract the 25 trillion cubic feet of gas that might lie under the Bowland shale in Lancashire would require around 3,000 separate well sites to be drilled. If each well produces 5 billion cubic feet of greenhouse gas? The numbers become unthinkable.

One estimate even suggests that up to 6% of the land surface of the UK would become a fracking site. The loss of vegetation from the land clearing required for this scale of operation would by itself create an increase in CO_2 emissions – but that's just for starters.

Fracking requires vast amounts of water to break apart the shale deposits and release the gas. A generally accepted estimate is four billion gallons for a single job and in operations carried out to date, each well has been fracked up to eight times. This water has to be transported to the site by huge tanker trucks, along with all the equipment, and materials needed to complete the job. The end result is a staggering 4,000 heavy vehicle movements over the production process for a single well site.

The massive use of water in drought-hit parts of the United States is increasing pressure on already depleted local supplies. According to the study by the Ceres green investors network, some of the most drought-affected areas of the country are also those experiencing increased shale gas development. About three-quarters of about 40,000 oil and shale gas wells drilled in the US since 2011 were in areas already facing water scarcity. More than half of the new wells were in areas of the US experiencing high water stress due to drought conditions, while nearly half were in areas already experiencing extremely high water stress.

Mindy Lubber, president of Ceres green investors' network, said, "Hydraulic fracturing is increasing competitive pressures for water in some of the country's most water-stressed and drought-ridden regions." Texas, one of the areas in the US experiencing the worst drought conditions, also has the highest concentration of fracking activity

Fracking also has potential for contaminating groundwater in several ways. The first is through the release of fracking fluid from horizontal wells, mixed with naturally occurring contaminants released by fracking from the shale, migrating upwards into aquifers through the intervening rock. Hydraulic fracturing increases the permeability of shale beds, creating new flow paths and enhancing natural flow paths for gas leakage into aquifers.

To understand why water contamination is inevitable from hydraulic fracturing, you need to understand the basic construction of a shale gas well. The drill bit, in its passage down into the earth, is encased in a basic steel pipe. This is intended to isolate the drill from the surrounding geology as it goes down, including through any aquifer near the surface that supplies drinking water to humans and animals.

The gap between the casing and the walls of the shaft is filled out by concrete – or at least it is supposed to be. Professor Tony Ingraffea (formerly an employee of the US gas industry) has spelled out critical faults that can occur with this kind of construction[14]. These include corrosion or damage to the steel casing – as happened at Cuadrilla's first Lancashire site – or if there is an unexpected naturally occurring tract of gas close to the shaft, or the concrete fails. These problems can amount to a loss of integrity of the well, in which case methane gas migrates from the shaft into the surrounding rocks. This will ultimately flow into the aquifer, irreversibly contaminating an entire drinking water source. Data from the oil and gas industry itself acknowledges that the majority of drilled oil and gas wells will ultimately be leaking.[15] Studies have found that methane concentrations in groundwater tend to increase in closer proximity to a gas well.[16]

In August 2012, a report from Stony Brook University, New York found that wastewater from fracking posed substantial risks of river and other water pollution. It found "even in a best case scenario, an individual well would potentially release at least 200 cu.m. of contaminated fluids." Disposing of fracking fluids in industrial wastewater treatment facilities could lead to elevated pollution levels in rivers and streams because they are not designed to handle wastewater containing high concentrations of salts or radioactivity.

As Richard Mills points out in *Fracking – an unconventional poisoning,* the water that flows out of the well can contain a variety of toxic and carcinogenic substances, most of which are not contained in the fracturing additives. This is because chemicals and minerals present in the shale zone may be released during the hydraulic fracturing process.

"The recovered waste fluid – water contaminated with chemicals and anything that water has come in contact with, meaning heavy metals and minerals – is often left in open air pits to evaporate, releasing harmful volatile organic compounds into the atmosphere, creating contaminated air, acid rain, and ground level ozone. Some of the recovered waste water is injected deep underground in oil and gas waste wells or even in saline aquifers. There are serious concerns about the ability of these caverns and aquifers to handle the increased pressure and in the US, evidence is showing that deep-well injecting is linked to the occurrence of earthquakes. According to the industry's own numbers just 60-70% of the fracturing fluid is recovered, the remaining 30 to 40% of the toxic fluid stays in the ground and is not biodegradable."[17]

In 2013 scientists from Duke University uncovered high levels of radioactivity at a shale gas waste disposal site in Pennsylvania. Water discharged from Josephine Brine Treatment Facility into Blacklick Creek, which feeds into a water source for west Pennsylvania cities, including Pittsburgh. Scientists took samples upstream and downstream over a two-year period and found elevated levels of chloride and bromide, combined with strontium, radium, oxygen, and hydrogen isotopic compositions, were present in the shale wastewaters.

CASE STUDY

"Wells are counted by the hundreds of thousands in the US and Canada, millions have been fracked worldwide and we're drilling hundreds more per day. Each and every one a potential ticking time bomb of human cancers and mutation. In the end, when the shale boom goes bust, and it'll be much sooner than most think, we'll have to live with what's been done to our environment. In a few short years will we be able to rationalize, to justify the short term benefits from poisoning our most precious resource, our fresh water?" Richard Mills, *Fracking – an unconventional poisoning*

The Environment Agency warned the UK government about water contamination but are working on a watered down version of their actual view. Their initial position, leaked to Greenpeace[18], was unequivocal: 'The Environment Agency would not allow hydraulic fracking to take place in an area where there are aquifers used to supply drinking water. If there were sensitive groundwaters present in an area where a company wanted to carry out hydraulic fracturing, we would object ... and refuse to grant an environmental permit." But the EA's head of climate change, Martin Diaper, suggested the wording be changed in order not to provide "too stark" a view of their position.

"I am a bit concerned that the two sentences ... provided a too stark view of our position of where we would or would not be happy with shale gas developments in relation to potable ground water aquifers. We take a risk-based approach to permitting ... Can I ask that you do not use the two sentences ... while we finesse them."

To sum up, shale gas is bad in terms of greenhouse gas emissions, bad in what it does to the environment and disastrous in its impact on water. As Kevin Anderson from the Tyndall Centre says:

> "Shale gas is the same as natural gas – it is a high-carbon fuel, with around 75% of its mass made of carbon. For the UK and other wealthy nations, shale gas cannot be a transition fuel to a low-carbon future. Anyone who says differently does not understand our explicit international commitments under the Copenhagen Accord, the Cancun Agreements – or, alternatively, is bad at maths.
>
> In the UK and globally, we are now reaping the reward of a decade of hypocrisy and self-delusion on climate change. We pretend we are doing something ourselves, whilst blaming others for rising emissions. The truth is out – it is a tragedy of the commons par excellence – we are all to blame and we have left it too late for a technical fix. We are heading towards a global temperature rise of 4ºC to 6ºC this century; if we want to get off this trajectory, shale gas needs to stay in the ground..."

6

It's a socio-eco crisis – time to act

Fracking for gas and oil, and the widespread opposition to it, contains all the elements of today's extreme economic, ecological, social and political problems, which share the same underlying cause – capitalism's systemic crisis.

It is an eco-social crisis, where people with no democratic control over their destiny in their own home, village, town or city are viewed by powerful élites as little more than a blockage to developments they believe are essential for the maintenance of the capitalist, growth-driven economy. Fracking makes sense only to those who are driven by profit and economic growth.

To the shareholders who own Cuadrilla, Anadarko, BP, Halliburton, Shell, ExxonMobil, Hutton Energy, Encana and PKN Orlen and to those who subscribe to the model of profit-driven growth, the economic argument is unassailable. They will employ any means necessary to maintain the process. The only kind of sustainability they understand is one where their profits grow.

Communities are regarded as obstacles to be swept aside. In October 2013, an editorial in *The Economist* urged ministers to forget about using tax breaks or spending money to encourage people to remain, or businesses to invest, in cities and towns like Hull and Burnley. This only diverted them away from areas where "they would be more successful". Governments should not try to rescue failing towns, but rather encourage the people who live in them to escape, said the business magazine.

In a debate on fracking in the House of Lords, Lord Howell (George Osborne's father-in-law) let the cat out of the bag about how the ruling élite see Britain's poorest areas. He started off by saying that whilst fracking might not be acceptable in the "beautiful" South East (he meant Balcombe) it would be fine for the "desolate North East". Then he made it even worse in his so-called apology by saying he actually meant the "unloved North West". Capitalism's view of the future is that former industrial zones are not for people but for fracking.

The accelerating crisis we are living through today has its origins in the rapid growth from the late 1970s onwards. A collapse of the post-war system of fixed currencies, capital controls, tariffs and tough regulations had shown that the forces of capitalism could not be contained within national boundaries. The old order was destroyed and replaced by corporate-driven globalisation. An entirely new global economy emerged, alongside a financial system that traded 24 hours a day and paid no heed to national borders or governments.

Giant global corporations were created, funded by credit. Their profits were extracted from labour cheapened at home and workers on subsistence wages in the so-called emerging economies. The rapid increase in production propelled demand for coal, oil and gas as energy sources – and led to the steep upwards curve in carbon emissions. The 5th IPCC report (2013) reiterates the message that climate change is the result of human activity – or more precisely 150 years of industrial capitalism.

When the conventional US oilfields began to run dry, interest in other sources led back to the Middle East, but as demand increased, and production headed towards its peak, world prices rose enough to stimulate interest in more expensive, and hazardous methods of extraction. Production and consumption of natural gas in the United States, the world's most significant capitalist economy, were in approximate balance up to 1986. Production then lagged consumption during the following 20 years; the deficit was made up largely by imports from Canada, delivered by pipelines. The situation changed dramatically in 2006 as companies using new drilling technologies moved aggressively to tap the vast supplies of previously inaccessible gas trapped in underground shale deposits. Natural gas extracted from such sources accounted for 10% of US production in 2007, and rose to 30% of production by 2010 – an enormous, swift change in the huge market.[19]

But, just as production volumes soared, the financial crash of 2008 and subsequent recession led to falling demand and saw prices plummet. The success of fracking in the United States, measured by the volume of gas and oil output is certainly impressive. But that success already brings new economic problems. The competitive rush to production combined with the crash and the muted nature of the economic recovery to which the low price of gas contributed, means that supply exceeds demand. And, as every classically trained economist can tell you, excess of supply over demand means prices must drop.

The other side to the story is that hydraulic fracturing combined with horizontal drilling is diabolically expensive and highly capital-intensive. Production from each well drops rapidly during the first year. So more pumping of the toxic mixture of water, chemicals and sand is required and the cost of maintaining the flow increases rapidly.

Continued low prices threaten the economic viability of the whole operation. Consequently, production has either to be greatly reduced to drive up the price and the benefits to the US economy greatly diminished or new infrastructure has to be developed to transport the gas and oil onto the world market where it is expected to attract a higher price.

The shale gas industry in the United States is heading for a shock. In fact, some observers have designated the process as a "shale gas bubble" waiting to burst. Two reports published in 2013 by the Post Carbon Institute (PCI) and the Energy Policy Forum (EPF) conclude that "the so-called shale revolution is nothing more than a bubble, driven by record levels of drilling, speculative lease and flip practices on the part of shale energy companies, fee-driven promotion by the same investment banks that fomented the housing bubble".

Fracking is just one of the contributors to the insatiable demand for more energy. Add in coal seam methane and deep sea drilling in the Mexican Gulf and the Arctic. In Australia and Indonesia, huge opencast mines scour the earth for coal to feed the commodity production carried out by global corporations in China. Some 70% of China's energy is from coal. In the 25 years to 2013, its annual consumption of coal has risen from 1 billion to 4 billion tons.

The ecological impact of all this is of little if any concern to the for-profit corporations whose sole interest is in maximising availability of the cheapest possible, highest concentrations of energy. The consequences are often disastrous.

Tepco's cost-minimising construction strategy and sidelining of maintenance of the Fukushima nuclear plant left it vulnerable to known risks. More than 90,000 people remain unable to return home. The plant itself is in an extremely vulnerable state. In October 2013, heavy rainfall led to water with high levels of the toxic isotope Strontium-90 overflowing containment barriers around water tanks.

BP and its supply companies ignored engineers' advice on measures needed to stabilise and secure its deepwater drilling. In 2010, the BP oil rig Deepwater Horizon exploded, killing 11 workers. Over 87 days, the damaged Macondo wellhead in the deep sea – around 5,000 feet down – leaked an estimated 4.9 million barrels of oil into the Gulf of Mexico, making it the largest ocean spill in history. While BP has abandoned its clean-up operations, local people are dealing with the consequences. A 40,000-pound tar mat was found in June 2013, part of the 2.7 million pounds of oiled material collected in the first six months of the year alone. It is difficult to say how much could be left for coming years.

The necessity for growth is a given, built-in to the system. For capital to keep expanding, as it must, everything becomes a candidate for exploitation in the commodity markets, sooner or later. That's the driving force behind the exploitation of shale gas. That's the motivation for cutting wage costs and cutting safety corners, even though the consequences can be deadly as Deepwater Horizon and Fukushima tragically demonstrated.

Yet the global economic crisis precipitated by the 2007-8 financial crash has shown to millions around the world that capitalism is not a sustainable system, socially or politically. For the majority, real incomes and the buying power they provide are declining fast as a result of the austerity policies supposedly designed to restore growth after the crash.

Contradictory forces are at work deepening the crisis. Pressure for growth continues, whilst its opposite – the necessity for elimination of surplus productive capacity intensifies. What will be done with the 60% of young people in Greece and Spain for whom there is and will be no work?

Humanity now has a clear choice. We can either sacrifice the conditions for life using all of the creative, scientific and technological ingenuity in the service of an economic system which destroys everything in its path. Or we can make an evolutionary leap in social organisation, shedding the old skin of capitalist social relations. Doing this will enable humanity to release the potential of collective, co-operative social relationships directed to producing what we need safely, using technologies which restore the planetary eco-system.

7

Frack capitalism to build a sustainable future

Slowing, then halting, global warming requires an immediate reduction in the quantity of fossil fuels burned. The major question facing humanity is: can we rely on the existing governing structures to make this happen? The evidence against is clear enough.

The 5th IPCC report (2013) shows that the world's "carbon budget" – the amount of greenhouse gas that can be emitted without exceeding 2ºC warming – could be used up entirely by 2040. The amount of CO_2 and other greenhouse gases trapped in the atmosphere is already enough to cause significant climate change, particularly more extreme weather, rising sea levels and increased ocean acidification.

The IPCC's first report was in 1990. At the summit that followed, governments adopted the Framework Convention which agreed that developed countries would stabilise their greenhouse gas emissions by the year 2000. That target, and every subsequent one, has been missed. Any reductions in Europe have been overwhelmed by growth in China, Brazil, India and Russia. As a result, in May 2013 the highest ever concentration of CO_2 was measured in the atmosphere, at over 400 parts per million.

The Department of Energy and Climate Change's "fact sheet" published in response to the 5th IPCC report (2013) highlights the problem. After summarising and accepting all the findings, it then skips straight to "What can I do?". It makes tackling climate change an individual responsibility and

43

says nothing about the responsibilities of government or industry.

The fact that the government is proposing to permit fracking and methane extraction across huge areas of the United Kingdom is the clearest expression yet of their refusal to act. It is not as if they don't know what they are doing. The Department of Energy and Climate Change itself commissioned a report to look at potential "lifecycle" carbon emissions from shale gas, including drilling, extracting and burning the gas. It showed that it would make no reduction in Britain's carbon emissions, and might well increase them, depending on what the gas replaces.

The report points out that the US switch to unconventional gas released huge quantities of cheap coal for export. Britain is one of the countries that has taken advantage of falling prices. A global fracking boom will ensure there is no incentive for China or others to seek alternatives to coal. A shale industry here would both increase global carbon emissions and make it impossible for the UK to meet its legally-binding targets under the Climate Change Act.

Can we look to the European Union where a new draft law on fracking will be published by the end of 2014? Britain is working with Poland and other Eastern European members to prevent strict controls, and a "light touch" approach is likely. Industry Commissioner Antonio Tajani sums up the reason, saying there will be a "systemic industrial massacre" unless energy costs are brought down to nearer US levels. "We need a new energy policy. We have to stop pretending, because we can't sacrifice Europe's industry for climate goals that are not realistic," he said. In the present system, growth will always trump tackling climate change in any government forum.

Pablo Solon was Bolivia's representative at the UN climate talks, including at Cancun where Bolivia was the only country that refused to sign the dirty deal that would inevitably lead to 4ºC of warming. He knows these talks inside out and concludes:

> "Climate negotiations make slow progress not because climate science is unconvincing or public awareness is lacking. Rather, élites and multinational corporations have captured the negotiations. Though diplomats from many countries understand that emissions of greenhouse gases remain too high because of consumerism and the pursuit of economic growth, few are willing to discuss changes to the economic system that drives climate change."

The alternative – a co-operative, not-for-profit economy

The only way forward is to end this system, and move to one that does not demand continuous growth in commodities and profits. It is well within our social, technological and organisational capacity to make the transition to a not-for-profit economy, producing only what people need to live well. This can only be achieved by co-operation. So the system's inbuilt priorities of competition and maximising shareholder value – enforced by law and driven on by political state structures – have to be replaced.

We can transform corporations into what Karl Marx called "a society of freely associated producers". These can operate locally, nationally and transnationally too. Worker-owners can set prices that incorporate high environmental standards and a social surplus to fund the common good.

The characteristics of the corporation operating this framework could be:
- co-operatively owned and run
- no shareholders and no profits – a social surplus to contribute to pensions, health and care
- functioning within a new kind of market that sets out to identify and meet need
- working closely with consumers to design products
- minimising inputs of raw materials and energy
- maximising product lifespan – produce for quality: design products that are "made to be made again"
- production as near to the point of consumption as possible
- share technology and research freely
- obliterate the concepts and principles of planned and perceived product obsolescence
- curtail the activities of the marketing and advertising industries.

Transition to a low-carbon energy future

The UK taxpayer currently subsidises the oil and gas industry by £280m per annum in tax breaks. This will increase dramatically if deep water oil and gas platforms and on-shore fracking come on line.

There is a £4bn per annum subsidy on direct consumption of petrol, gas and coal through a reduction in VAT from the standard 20% rate to 5% (this also applies to renewables). There is no technological or security of supply benefit from subsidising fossil fuel; future security lies in escaping its grip. Therefore any public subsidies must go to the development of renewable energy and a smart grid.

It is clear to everyone – even the ConDem government – that the market cannot make the transition to renewables or go on meeting demand (at prices people can afford). The government's proposed Electricity Market Reform

(EMR) admits it is a response to "market failure". However, all it does is more deeply embed the principle that taxpayers subsidise the profits of the Big Six energy companies, whilst they go on freely polluting.

The EMR says the government will allow the energy giants to charge artificially high prices to consumers for a fixed period – so-called contracts for difference – in return for new infrastructure, including nuclear power stations. This is the only way new nuclear or any other infrastructure can happen – none of the corporations wants to accept any risks or spend for the long-term. This policy was behind the outrageous £70 billion of taxpayers' money now pledged to the French and Chinese state-owned corporations that are to build and run a new nuclear power station at Hinkley Point, Somerset. This project is going ahead with no plan for how the nuclear waste will be cleaned up, not even how it will be stored for future generations to worry about.

End the power of the Big Six

The urgent first step would be to bring the energy corporations into this new not-for-profit model, taking national action through a People's Assembly, in order to protect fuel supplies whilst acting urgently to reduce greenhouse gas emissions and create a community-based energy model.

By putting energy under democratic control, unfair prices could be ended and fuel poverty eradicated virtually overnight. We would open the books of the energy companies to get at the truth of what is investment in new or improved infrastructure and what is simply a profit grab.

Start the transition

There is nothing technological preventing a transition to low-carbon energy. Offshore wind power alone could meet between 60% and 70% of projected European demand for electricity in 2020 and about 80% of projected demand by 2030, with reasonable investment by governments. Nine years have passed since the EU produced this figure, and governments did not act on it. In the same year, a report in *Scientific American* brought together a range of evidence to show how the world could become entirely de-carbonised also by 2030. All of this evidence has been ignored in energy planning.

The example of Germany shows that even in the present it is possible to significantly expand wind and solar. One of the projects funded by Obama's Green New Deal has started developing a salt water battery for storing variable power – a real breakthrough. But research and development is limited by an entirely extraneous factor – how quickly a given technology can produce dividends for shareholders. This "everything must be a commodity"

thinking underlies the "Green Capitalism" and "Eco-services" proposition, which is doomed to fail to protect the eco-system, but which even some conservationists have despairingly adopted as the only way forward. With this out of the picture we could set different priorities for energy generation:
- national, regional and local transition plans to phase out fossil fuels whilst protecting supply, but not supply for wasting
- co-operatively owned and run energy firms produce an energy efficiency plan to be democratically consulted on and agreed
- demand reduction – firms will share/work co-operatively with others in their industry/commercial sector to develop ways to reduce energy use
- building on existing examples, support communities to form local energy co-operatives
- electric power generated as close to the point of consumption as possible, a flexible grid
- local people's assemblies take a lead on a community energy plan to include insulation, energy saving and demand reduction; small scale renewables including combined heat and power, solar, wind, tidal, mini-hydro – the right local mix at an affordable price.

Ownership – the creation of a new global commons

Fossil fuels have played a major part in capitalism's progress to the present totalising global system. Without the properties of coal would capitalism be what it is today? That is one question, but the other question is – coal was there for ever and its flammable properties known. What brought about its central economic role? The role of fossil fuels as climate killers does not result from their inherent properties, but from the economic framework within which they are exploited – socially-created systems that are in direct conflict with natural, restorative systems.

The growth in extraction of fossil fuel in Britain went alongside the enclosure of common land and its transfer into private hands. This primary accumulation was the foundation for the industrial revolution. Land laws were framed in the late 18th and 19th century by a parliament that represented the growing power of industry and financial speculators. English Common Law was no match for them. Parliament actively set out to destroy commons and transfer land into private hands. Often commons were enclosed illegally and the people driven out; ownership was then confirmed post-facto.

Landowners quickly cashed in on coal deposits on land they now "owned", selling rights to industrial barons or sometimes exploiting it themselves. We urgently need new democratic institutions, with lawmaking powers, to underpin the creation of new commons. The New Commons will include all the resources of Mother Earth, and materials or knowledge extracted from these resources. It will not be possible to privately own these, or to copyright any natural product or process.

> **CASE STUDY:**
>
> It is reported that Scottish aristocrat and millionaire the Duke of Buccleuch is ready to cash in on the fracking boom at one of his Scottish estates near Canonbie in in the Scottish Borders. He will not be the first Scottish aristocrat to get richer from fossil fuels. The 2nd Marquis of Bute (1793 - 1848) owned thousands of acres in the Vale of Glamorgan and became not only a multi-millionaire coal baron, but also built Cardiff docks and charged big fees to coal exporters/iron ore importers.

Legal rights for Mother Earth

Law is one of the instruments humans have developed to regulate society. Under capitalism much of it is class law, protecting above all else the rights of the thieving minority over the working majority. But we can revitalise the principles of a law-governed society with a legal code aligned with natural law.

We need an internationally-accepted legal framework that recognises that there are natural limits that cannot be transgressed and that we live in an interdependent relationship with all of the species on planet earth. So the framework of law – from local People's Assemblies up to a democratically transformed World Trade Organisation and World Bank – can become the means of implementing the end of the alienated capitalist model and the start of a renewed relationship with nature.

> **CASE STUDY:**
>
> In October 2012, Bolivia passed the Law of Mother Earth, a bill of rights for the natural world including the rights to biodiversity, uncontaminated water and air, no genetically modified crops and no overdevelopment. It states: "The environmental functions and natural processes cannot be considered as commodities, but as sacred gifts from Mother Earth." There is an Ombudsman for Mother Earth, and a framework for the responsible use of Bolivia's vast mineral and hydrocarbon reserves.

An action plan for the UK - 2014

Local democracy
- form local People's Assemblies to debate and agree land use, economic and social strategies
- pursue community common land ownership, by using existing law, demanding new laws and by occupying spaces the community needs
- community groups and local charities sign a co-operation agreement, so they no longer compete against each other for funds or get involved in cost-cutting
- demand community-based decisions on how funds are spent, through the People's Assembly
- start to regenerate run-down areas through community-based activities
- demand a greater community say in what happens in schools to support young people's creativity and self-esteem
- liberate school pupils and young unemployed people to stop chasing dead end qualifications and jobs and instead be actively involving in running their community
- skill sharing between the generations.

CASE STUDY

Planning resistance to unconventional gas developments has made people start thinking about how they want their communities to be. The group Concerned Communities of Falkirk not only organised opposition to Dart Energy's plans for coal bed methane capture, they have proposed an alternative. Their Community Charter sets out the area's "cultural heritage" which is declared to be:

"The sum total of the local tangible and intangible assets we have collectively agreed to be fundamental to the health and well-being of our present and future generations. These constitute an inseparable ecological and socio-cultural fabric that sustains life, and which provides us with the solid foundations for building and celebrating our homes, families, community and legacy within a healthy, diverse, beautiful and safe natural environment. This is the basis of a true economy, one which returns to its root meaning (oikos - home, nomia - management)".

Energy transition
- consult widely on an energy transition plan to move from fossil fuels to renewables with a realistic timescale and a sense of urgency
- end fuel poverty by bringing energy companies into common ownership under democratic control
- set an end date for all public subsidies for nuclear and fossil fuel energy
- consult on transparent plans to safely store or recycle existing nuclear waste
- funds for community-based local energy programmes, with insulation for homes and small businesses and small-scale renewables
- invest in a decentralised national grid built around two-way flows and designed to work effectively with non-fossil resources
- redesign, renovate and construct all buildings with as many known features as possible to minimise and eliminate a need for access to fossil fuel generated energy, and to take account of the highly likely local effects on weather of global climate destabilisation (mainly flooding).

CASE STUDY:

Bristol Energy Cooperative (BEC) is a community-owned energy cooperative developing renewable energy and energy efficiency projects. The co-op has investor members who receive a return on investment from the energy the co-operative produces. The surplus is used to invest in further projects. In 2012 they raised £128,000 to install Solar PV panels on community buildings. The result is 250 panels generating 63kw. One can imagine a scenario where a firm invested its social surplus in a local energy company, ensuring a reasonable return to spend on social care, health and pensions for its employees.

Transport
Transport is responsible for more than a fifth of the UK's emissions, and the largest proportion of that comes from road transport. A new transport policy is needed:
- reduce travel, with people working at local centres, and thousands of new smaller local schools to minimise car and bus journeys
- bring rail and bus networks into not-for-profit common ownership under democratic control
- transport planners work to create integrated transport systems

- discourage car use in cities by park and ride, public transport and car exclusion zones
- support local car pools and car-sharing projects, cycle lanes and walking
- set upper limits on total flight miles and distribute them through an air miles system
- halt airport expansion and all airports to deliver environmental improvement plans.

CASE STUDY:

Communities in Cornwall are trying to tackle rural isolation with community-owned and run public transport. There are voluntary car schemes where people can organise a lift to a health appointment or other essential meeting; community mini-buses operating to timetables and run by volunteers; minibus hire for groups plus driver training; cheap mopeds or bikes to help young people get to work, school or college – training and safety gear included. A huge number of community groups, charities, churches and individuals are involved and most are volunteers.

The battle of ideas

Capitalism is underpinned not only by a state-backed framework of laws but also by an ideology. This ideological paradigm is spread through the education system and the media, through culture and even through competitive sport. It advances a model of humanity as the controller of nature, recognising no external limits to what we can do. As a columnist in *The Telegraph* wrote about the US oil and gas boom: "The global energy crisis has been postponed yet again. It's time to forget about 'peak oil'. Human ingenuity has saved the day."

But whilst humans can control aspects of the natural world, and can even destroy whole species and ecosystems, they are nonetheless wholly part of nature and subject to its laws. They are only one of many co-existing species that make up the biosphere. It is true that every life form is in a life and death struggle to get from its environment what it needs in order to survive and that there is competition for resources between species. The first aspect of our existence – our unity with nature – is absolute and infinite. The second – the conflict – is relative and finite. In capitalist ideology, this relationship is stood on its head. The struggle to wrest from nature what we need for survival is elevated to an absolute. The unity and interdependence, and the finite nature of ecosystems, is ignored or turned into a pleasant abstraction to be contemplated when on holiday somewhere green and pleasant.

However this unity is no empty abstraction, but the objective and concrete reality of our existence. Increasingly, science is investigating the interconnectedness of life, from the smallest particle to the most developed species. Evolutionary biologists are developing a fresh approach which they title co-evolution, showing how eco-systems grow together as a unique whole of diverse species, mutually conditioning each others' evolution. The key to evolutionary success is diversity – not the eradication of thousands of species in the process of extracting raw materials.

Under capitalism even our human relations are profoundly alienated and commodified, clearly a major source of unhappiness and mental ill health. Yet we bravely continue to struggle against these damaged relations because of our inner feelings, our intuition and sense of what is right and wrong, and our ability to care for others. The challenge is how to go beyond individual caring actions, or even active resistance to capitalism, to overthrowing its hegemony and ending its rule.

As the eco-social crisis deepens, increasing masses of people are aware there has to be an alternative. Masses of people are resisting, from the Arab Spring, to Occupy, to opposing fracking in their area, to cutting their own commodity consumption. People across the globe are ready for alternatives but still struggle to break free of the idea that all we can do is protest against a system that continues to claim a natural right to rule, whereas what we need is ideas for fundamental change.

People's Assemblies

In order to address this conundrum and create the framework for change, we need a new framework for independent self-government and democracy. The People's Assembly model offers a new kind of participative democracy, where people have the authority to make plans for their community and implement them. It must take the place of the capitalist state system that upholds and advances the status quo, at whatever cost to humanity.

There will be no professional politicians. Groups can send delegates to the assembly – young or old, in work or unemployed, in the public or private sectors, trade unionists, women, students, minorities and community groups. But people will always be able to represent their own individual views on any topic and to be involved in any area of work or decision-making. Their confidence and self-reliance will increase exponentially.

A network of People's Assemblies can get to grips with the eco-crisis in a way that no local or national government has been able to. The Assemblies will think about how local land and resources are used, and work with neighbouring Assemblies on shared issues. They will develop energy plans and work with local farmers and other producers to promote local consumption and low energy organic farming.

Democratic global governance

The same vision can underpin a global People's Assembly. It will be complex, no doubt with strongly differing views, but at least it can make progress in a way the United Nations and other global forums have entirely failed to do. Unlike at present, ferocious competition between corporations and nations will not act as a total block on an internationally binding agreement to reduce greenhouse gas emissions. They can assess the contribution geo-engineering strategies could make in preventing runaway climate change in the short term against the harm they may do to planetary ecosystems.

The global assembly can plan to mitigate the climate impacts that are now inevitable and to support people and peoples to adapt. Instead of ignoring the scientific assessments and other studies coming from the IPCC and others, they would draw on all the expertise represented by climate scientists, food, agriculture and health experts, ecologists, philosophers, historians, religious leaders, artists and even poets.

Within this framework, nations can support each others' development towards self-government and economic independence. The road to ending exploitation and corruption and having equality for all in every country may be long, but a global People's Assembly that does not represent corporate or colonialist interests can help map it out.

Build Peoples Assemblies now – the forum for transition

The great strength of the People's Assembly concept is that we can begin to create them right now – they can start off as the forum for fighting for the change, and as the embryo or test-bed for the new form of democracy, which will be our future.

People's Assemblies are a democratic alternative to a failed, undemocratic political system. Technology means that in one way or another, through representing themselves or designating people to represent them, the people of the planet could participate in all the assemblies at local, regional, national and in the future even global level. They can learn from each other, and make sure they get unbiased expert opinion to inform their decisions. They can adopt the principles of non-violent communication to make sure that sectarian and self-interested conflicts cannot dominate.

New ways of consulting using film, theatre, music and other participatory principles, will replace tedious focus groups and "polling", and add to the principles of one person one vote. Instead of a bloated bureaucracy serving its own interests and the interests of the élite, we'll be having to find new ways to cope with an overwhelming outbreak of democracy.

A people's inquiry
Our future *beyond* capitalism
**Running now on A World to Win's networked forum
Join today and contribute to a plan for change**

This People's Inquiry offers a chance to share ideas about the related crises facing humanity. It is open to individuals, campaign groups, trade unions, academics and students. Working together, we can deepen our knowledge and work on solutions.

Climate change has passed a tipping point and extreme weather is just one of the results. The global economic and financial system has entered a renewed, deeper crisis. Anger at the way political systems are corrupted by corporations and lobbyists is at boiling point.

There is worldwide opposition and resistance, but as yet no shared strategy for getting beyond capitalism. Reaching the point where we achieve such a strategy is the main purpose of the inquiry. There are six study areas:

- The ecosystem, including climate change and species loss
- Global economy and finance, where the 1% rule over the 99%
- The state, democracy and social rights, like health and housing
- Ideology and philosophy – dialectics of liberation
- Culture, education and sport – how they can help set us free
- Networks/organisations/strategies for revolutionary change

It is a 4-stage process up to a final report in October 2014 and beyond:
1. Gathering evidence through the network group
2. Face-to-face meetings across the UK, with on-line participation
3. Working groups to collaborate on the contents of a draft final report that maps out a way forward
4. Plan to implement solutions

**You can register today to take part at:
aworldtowin.ning.com**

Notes

Ch 1
'A Retrospective Review of Shale Gas Development in the United States' http://www.rff.org/RFF/Documents/RFF-DP-13-12.pdf
2. http://www.huffingtonpost.co.uk/2013/06/28/power-blackout-ofgem-energy-electricity-national-grid-_n_3514405.html
3. Drill Baby Drill, David J Hughes http://www.postcarbon.org/reports/DBD-report-FINAL.pdf
4. http://www.rtcc.org/2013/10/03/bloomberg-fracking-unlikely-to-bring-down-uk-energy-bills/#sthash.gZaMhqXK.dpuf
5. EIA Annual Energy Outlook 2012 http://www.eia.gov/forecasts/aeo/pdf/0383%282012%29.pdf
6. http://www.theoildrum.com/node/10088
7. Technically Recoverable Shale Oil and Shale Gas Resources: An Assessment of 137 Shale Formations in 41 Countries Outside the United States http://www.eia.gov/analysis/studies/worldshalegas/pdf/overview.pdf
Ch 2
8. http://www.theguardian.com/environment/2013/sep/11/balcombe-residents-experience-poor-fracking-regulation
Ch 3
9. http://www.opendemocracy.net/civilresistance/philippe-duhamel/civil-resistance-as-deterrent-to-fracking-part-one-they-shale-not-0
Ch 4
10. http://www.independent.co.uk/news/uk/politics/how-lobbying-works--and-why-this-bill-wont-change-a-fracking-thing-8773474.
11. http://www.countercurrents.org/todhunter041013.htm
12. http://www.s2bnetwork.org/fileadmin/dateien/downloads/Brave_New_Atlantic_Partnership.pdf
13. http://www.nytimes.com/2013/10/19/world/europe/lobbying-bonanza-as-firms-try-to-influence-european-union.html
Ch 5
14. http://www.youtube.com/watch?v=fjaRwh4xRiM
15. http://www.slb.com/~/media/Files/resources/oilfield_review/ors03/aut03/p62_76.pdf
16. http://www.pnas.org/content/early/2011/05/02/1100682108.full.pdf+html
17. "Fracking – an unconventional poisoning". Richard Mills, http://aheadoftheherd.com/Newsletter/2013/Fracking-An-Unconventional-Poisoning3.html
18. http://www.greenpeace.org.uk/newsdesk/energy/investigations/foi-documents-reveal-confusion-over-shale-regulation
http://www.ceres.org/issues/water/shale-energy/shale-and-water-maps/hydraulic-fracturing-water-stress-water-demand-by-the-number
Ch 6
19. http://www.kpmg.com/Global/en/IssuesAndInsights/ArticlesPublications/shale-gas/Documents/cee-shale-gas-2.pdf

Unfinished Business
the miners' strike for jobs 1984-5

By Peter Arkell & Ray Rising | 96 pages & 80 photos: £7.99
Now that the economic and financial crash is joined by a deep political crisis within the British state, the opportunities are present for picking up where the miners were forced to leave off. Our challenge today is to build a movement that will complete the unfinished business of the miners' strike.

Unmasking the State
a rough guide to real democracy

by Paul Feldman | 88 pages: £3.99 & free pdf
Analyses the historical origins of the contemporary British capitalist state and the long struggle for democracy and political rights, from the Levellers to the Chartists and beyond. Describes the changes to the state under globalisation and how representative democracy has been undermined. Makes a series of proposals for a new, transitional state to extend democracy into workplaces and society as a whole.

A House of Cards
from fantasy finance to global crash

by Gerry Gold and Paul Feldman | 88 pages: free pdf
Your guide to understanding the crisis that is sweeping through the global financial system and what it means for ordinary people.

Running a Temperature
an action plan for the eco-crisis

By Penny Cole & Philip Wade | 64 pages: free pdf
A punchy analysis of the underlying causes of the destruction of the planet's eco-systems.

A World to Win
A rough guide to a future without global capitalism

By Paul Feldman & Corinna Lotz | 370 pages: £9.99 & free pdf
Our proposals seek to realise the potential that already exists through the creation of a new social, economic, political and cultural framework based on co-operation, co-ownership and self-management.

Available online www.aworldtowin.net